putting a price on sustainability

BRE
Centre for
Sustainable
Construction

Cyril Sweett
Sustainability and
Cost Consulting Teams

bretrust

This work has been funded by BRE Trust. Any views expressed are not necessarily those of BRE Trust. While every effort is made to ensure the accuracy and quality of information and guidance when it is first published, BRE Trust can take no responsibility for the subsequent use of this information, nor for any errors or omissions it may contain.

The mission of BRE Trust is 'Through education and research to promote and support excellence and innovation in the built environment for the benefit of all'. Through its research programmes the Trust aims to achieve:
- a higher quality built environment
- built facilities that offer improved functionality and value for money
- a more efficient and sustainable construction sector, with
- a higher level of innovative practice.

A further aim of the Trust is to stimulate debate on challenges and opportunities in the built environment.

BRE Trust
Garston, Watford, Herts WD25 9XX
Tel: 01923 664598
Email: secretary@bretrust.co.uk
www.bretrust.org.uk

BRE Trust and BRE publications are available from:
www.brebookshop.com
or
IHS Rapidoc (BRE Bookshop)
Willoughby Road
Bracknell RG12 8DW
Tel: 01344 404407
Fax: 01344 714440
Email: brebookshop@ihsrapidoc.com

Published by BRE Bookshop for BRE Trust

Requests to copy any part of this publication should be made to:
BRE Bookshop
Garston, Watford, Herts WD25 9XX
Tel: 01923 664761
Email: brebookshop@emap.com

FB10
© Copyright BRE Trust 2005
First published 2005
ISBN 1 86081 750 5

BRE
Centre for Sustainable Construction
Garston, Watford WD25 9XX
Tel: +44 (0)1923 664462
Email: breeam@bre.co.uk
Web: www.bre.co.uk

Cyril Sweett
60 Grays Inn Road
London WC1X 8AQ
Tel: +44 (0)20 7061 9000
Email: sustainability@cyrilsweett.com
Web: www.cyrilsweett.com

contents

summary

One of the principal barriers to the wider adoption of more sustainable design and construction solutions is the perception that they incur substantial additional costs. Evidence collected by BRE and Cyril Sweett contradicts this assumption. This report identifies the costs associated with a range of sustainable solutions for different building types, demonstrating that significant improvements in the sustainability performance of a building can be achieved at very little additional cost. In addition, this report also demonstrates that more sustainable buildings can offer major life-cycle cost benefits.

A house

An EcoHomes Good rating can be achieved for an additional capital cost of between 0.3% and 0.9%. Achieving a Very Good rating can be achieved with an additional cost of between 1.3% and 3.1% and an Excellent rating for between 4.2% and 6.9%, for a range of locations.

A naturally ventilated office

A BREEAM Good rating can be achieved for a *saving* of between 0.3% and 0.4% of the capital cost. A Very Good rating can be achieved for between a cost saving of 0.4% and an additional cost of 2% and an Excellent rating for an additional cost of between 2.5% and 3.4%, for a range of locations.

An air-conditioned office

A BREEAM Good rating can be achieved for an additional cost of between 0% and 0.2% of the capital cost. A Very Good rating can be achieved with an additional cost of between 0.1% and 5.7% and an Excellent rating for between 3.3% and 7.0%, for a range of locations.

A PFI-procured health centre

The base case health centre already achieves a Good rating. A Very Good rating can be achieved at no additional cost and an Excellent rating for an additional cost of between 0.6% and 1.9%, for a range of locations.

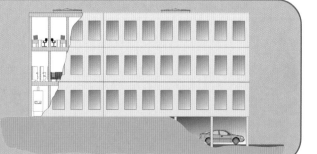

introduction

Sustainability — at what cost?

The drive for more sustainable development is one of the defining issues of the early 21st Century. It is often said that the costs of today's lifestyles are such that future generations will pay a high price through reduced environmental quality and living standards. However, it is also perceived that the short-term costs of more sustainable practices are too high to justify their application in a competitive property market.

Despite substantial advances in best practice, there is a lag in the application of more sustainable solutions that improve building performance beyond that required by Building Regulations. There are many reasons for this, not least a perceived lack of client/customer demand; however, one of the most often cited is that more sustainable alternatives are prohibitively expensive. Typically, cost consultants can add a significant margin of as much as 10% to capital costs to allow for more sustainable solutions.

This study is one of the first to examine this assumption in detail, providing real cost data for a broad range of sustainability technologies and design solutions. The results show that major performance improvements can be achieved cheaply and even at no cost at all. Reaching the highest standards of current practice does incur some cost premium, but this research shows that careful consideration of designs and specification at an early design stage can minimise these cost premiums compared with a more ad hoc approach.

This view is supported by the Sustainable Buildings Task Group in their report 'Better buildings — Better lives' published in May 2004.

Key drivers

Often the most powerful and direct driver for addressing sustainability is that the client, funder or planning authority has made it a key project requirement! For example:
- The Housing Corporation requires an EcoHomes 'Good' rating on any scheme they fund.
- English Partnerships require partner developers to achieve a minimum BREEAM/EcoHomes 'Very Good' rating.

- Public sector contractors should achieve a BREEAM Excellent rating for all new buildings and their proposals are often critically evaluated according to the extent to which they address sustainability issues.
- Many high profile private developers and landowners are seeking higher standards of sustainability performance from their partners.
- Investors are also becoming increasingly interested in sustainability and are engaging with major property industry partners to understand how they contribute to the wider sustainability agenda. A recent survey supported by the investment industry and WWF (World Wildlife Fund) listed the sustainability credentials of leading housing developers, indicating that investors are now considering sustainability performance in their investment criteria.

There is significant competitive advantage to be gained from actively addressing the sustainability agenda. This includes a need to understand the financial and programming implications of committing to BREEAM and other sustainability issues.

In addition to industry requirements, other drivers include:
- reduced running costs,
- improved living and working environments, and
- market differentiation.

Aims of this report

By identifying the real costs of sustainable solutions and thereby tackling a key barrier to the industry in advancing the sustainability agenda, this report:
- informs clients, designers and cost consultants of the true financial impact of environmental design options,
- provides broad cost benchmarks for achieving defined environmental performance standards for different building types,
- considers the cost implications of different solutions to determine 'quick win' options,
- highlights whole life cost implications in terms of energy and water consumption.

The costs of achieving enhanced and exemplar environmental performance for the following four types of building were investigated:
● a house,
● a naturally ventilated office,
● an air-conditioned office, and
● a healthcare centre.

These buildings were chosen to represent typical industry projects in the UK. BRE's suite of BREEAM tools was used to determine benchmarks of environmental performance.

The capital costs and life-cycle implications of each design, management or specification option was assessed and compared with a Building Regulations compliant standard (except in the case of the health centre where industry standards were higher). The capital cost quotes include prelims, overheads, profits and contingencies. The aspects of life-cycle performance presented are the savings in energy and water costs. Other aspects of life-cycle costs were not considered as they relate to the specifics of the site, the building occupier and maintenance methods employed.

BREEAM was selected because it provides a nationally recognised system for objectively assessing the environmental performance of a range of building types. The first three case studies were measured using standard versions of BREEAM, whereas the healthcare centre was assessed using Bespoke BREEAM. BREEAM measures the environmental performance of buildings by awarding credits for achieving a broad range of environmental standards and levels of performance. Each credit is weighted according to its importance and the resulting points are summed to give a total BREEAM score and rating (see the *Appendix* for further details).

By using BREEAM credits as a measure of environmental enhancement, it was possible to identify the additional costs associated with each incremental improvement in performance and thereby identify both the quick wins and the initiatives that need to be built into a project at the concept stage. The costs reflect items implemented at design stage and not retrofitted. The most cost-effective options were favoured.

In the BREEAM schemes, several credits relate to the site. These include proximity to local amenities and public transport, existing ecological value and whether the site has previously been built upon. In this study, three location scenarios were assessed:
● Poor location (where no location credits are achievable),
● Typical location (where a selection of credits are achievable such as a brownfield site with some access to local amenities and public transport, based on an edge of town location),
● Good location (where all location credits are achievable).

The results for each case study are presented in a summary graph and show how location is an important influence on the cost of achieving a high environmental rating.

Some credits could not be assessed in this study, as the costs were deemed to be too site-dependent. These include the cost of improving the ecological value of the site, planning for the long-term impact on biodiversity and the use of recycled aggregate.

As these credits could not be assessed, more innovative and potentially costly technologies such as rainwater harvesting and photovoltaic panels were incorporated. These innovative technologies may not represent the most cost-effective means of achieving a BREEAM Excellent. Depending on the site, achieving the site-dependent credits could provide a more cost-effective way to increase the BREEAM rating than some of the more innovative measures identified and costed in this study.

While this study focuses on new build projects, many of the costs are also applicable to building refurbishments.

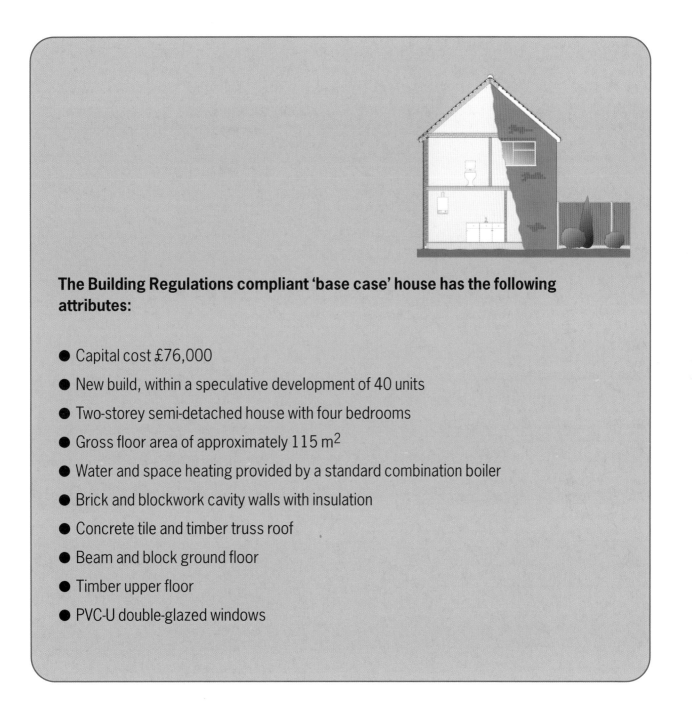

The Building Regulations compliant 'base case' house has the following attributes:

- Capital cost £76,000
- New build, within a speculative development of 40 units
- Two-storey semi-detached house with four bedrooms
- Gross floor area of approximately 115 m^2
- Water and space heating provided by a standard combination boiler
- Brick and blockwork cavity walls with insulation
- Concrete tile and timber truss roof
- Beam and block ground floor
- Timber upper floor
- PVC-U double-glazed windows

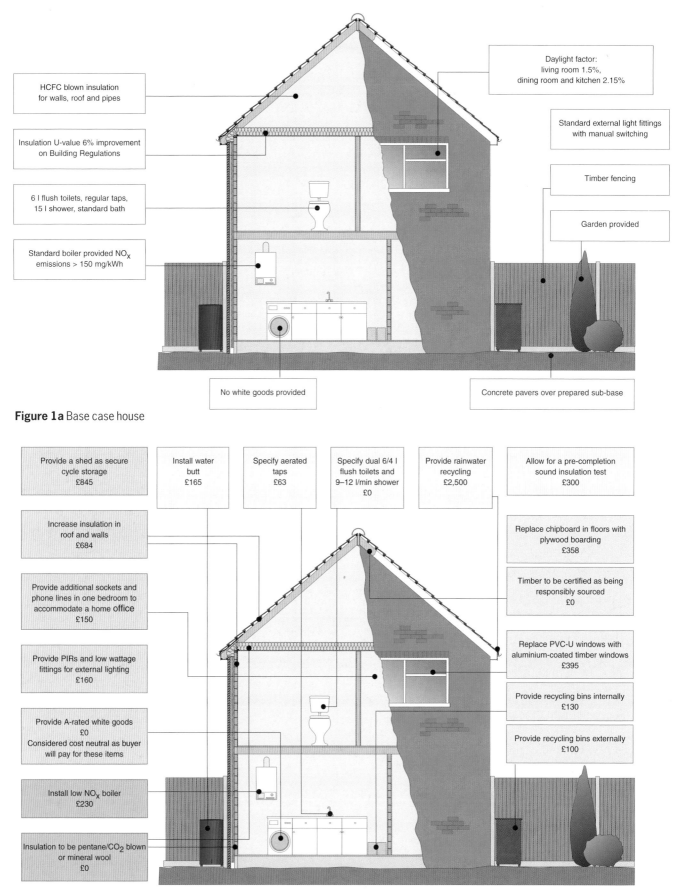

HCFC blown insulation
for walls, roof and pipes

Insulation U-value 6% improvement
on Building Regulations

6 l flush toilets, regular taps,
15 l shower, standard bath

Standard boiler provided NO_x
emissions > 150 mg/kWh

Daylight factor:
living room 1.5%,
dining room and kitchen 2.15%

Standard external light fittings
with manual switching

Timber fencing

Garden provided

No white goods provided

Concrete pavers over prepared sub-base

Figure 1a Base case house

Provide a shed as secure
cycle storage
£845

Install water
butt
£165

Specify aerated
taps
£63

Specify dual 6/4 l
flush toilets and
9–12 l/min shower
£0

Provide rainwater
recycling
£2,500

Allow for a pre-completion
sound insulation test
£300

Increase insulation in
roof and walls
£684

Provide additional sockets and
phone lines in one bedroom to
accommodate a home office
£150

Provide PIRs and low wattage
fittings for external lighting
£160

Provide A-rated white goods
£0
Considered cost neutral as buyer
will pay for these items

Install low NO_x boiler
£230

Insulation to be pentane/CO_2 blown
or mineral wool
£0

Replace chipboard in floors with
plywood boarding
£358

Timber to be certified as being
responsibly sourced
£0

Replace PVC-U windows with
aluminium-coated timber windows
£395

Provide recycling bins internally
£130

Provide recycling bins externally
£100

Figure 1b Additional capital costs for implementing more sustainable solutions (costs are for items implemented at the design stage, not retrofitted)

The environmental performance of the house was assessed using EcoHomes 2003 (see Appendix A). Figure 1b illustrates the environmental improvement measures assessed for EcoHomes compliance and the additional costs associated with them compared with the base case house. Table 1a lists the environmental improvement measures in order of best value for each of the environmental categories.

The % increases in capital costs shown in Table 1b have been calculated by applying selections of the measures listed in Table 1a to the base case house. The cheapest means of achieving the required rating was favoured.

Table 1a Additional capital costs for sustainable solutions along with the EcoHomes points achievable. The measures are presented in order of best value

Description	Cost (£)	EcoHomes points achieved[1]
Reducing CO$_2$ emissions from transport and operational energy		
Provide 'A' rated white goods	0	3.81
Provide PIRs and low wattage fittings for external lighting	160	2.14
Provide additional sockets and phone lines in one bedroom to accommodate a home office	150	1.07
Increase insulation in roof and walls	684	4.29
Provide a shed as secure cycle storage	845	2.14

Description	Cost (£)	EcoHomes points achieved[1]
Reducing mains water consumption		
Specify dual 6/4 litre flush toilets, aerated taps and 9–12 l/min shower	63	1.67
Install water butt	165	1.67
Provide rainwater recycling	2,500	3.33

Description	Cost (£)	EcoHomes points achieved[1]
Reducing the impact of materials used		
Timber to be certified as being responsibly sourced	0	4.35
Provide recycling bins internally and externally	230	2.90
Replace chipboard in floors with plywood boarding	358	1.45
Replace PVC-U windows with aluminium-coated timber windows	395	0.97

Description	Cost (£)	EcoHomes points achieved[1]
Reducing pollutants harmful to the atmosphere		
Insulation to be pentane/CO$_2$ blown or mineral wool	0	4.29
Install low NO$_x$ boiler	230	7.50

Description	Cost (£)	EcoHomes points achieved[1]
Improving the indoor environment		
Allowance for a pre-completion sound insulation test	300	3.75

---

[1] For a description of EcoHomes points, credits and scoring see the Appendix on page 21.

Table 1b   Increases in capital costs to achieve Good, Very Good and Excellent EcoHomes ratings in three locations

Location[2]	EcoHomes score and rating for the base case house	% increase in capital cost to achieve a Pass/Good/Very Good/Excellent			
		Pass	**Good**	**Very Good**	**Excellent**
Poor	22.1 (Unclassified)	0.1	0.9	3.1	—
Typical	27.6 (Unclassified)	0	0.4	1.7	6.9
Good	29.7 (Unclassified)	0	0.3	1.3	4.2

A good location can achieve an Excellent for as little as 4.2% extra on base capital costs. Even a location where a limited number of location credits are achievable, can achieve an Excellent rating for less than 7% extra on capital costs. To achieve an Excellent rating in a poor location it would also be necessary to consider the selection of site-related credits excluded from this study (see 'Our approach' on page 2) as a few more credits are required to reach the target level. A typical or good location can achieve a Pass for no additional cost. Figure 1c illustrates these results graphically.

As with all the graphs in this study, sometimes the lines cross the rating boundaries at a lower % than is given in the tables. Costs may be higher in the tables because it is not always possible to obtain the exact score for each rating boundary. A single improvement may increase the score from below 70% to 71% and thus the graph of cost versus score will indicate a lower theoretical cost for the required 70% boundary.

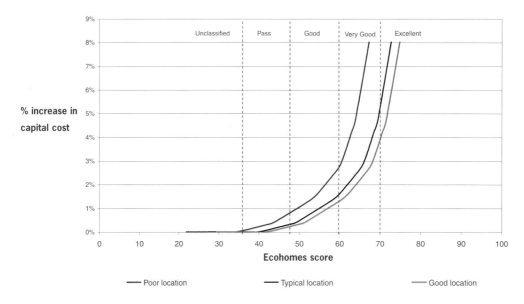

Figure 1c Cost versus environmental performance for the house

Whole life implications

The energy- and water-saving measures identified for the house provide predicted in-use cost savings of 6% and 40% respectively, throughout the life of the building.

---

2 For description of the three locations see 'Our appoach' on page 2.

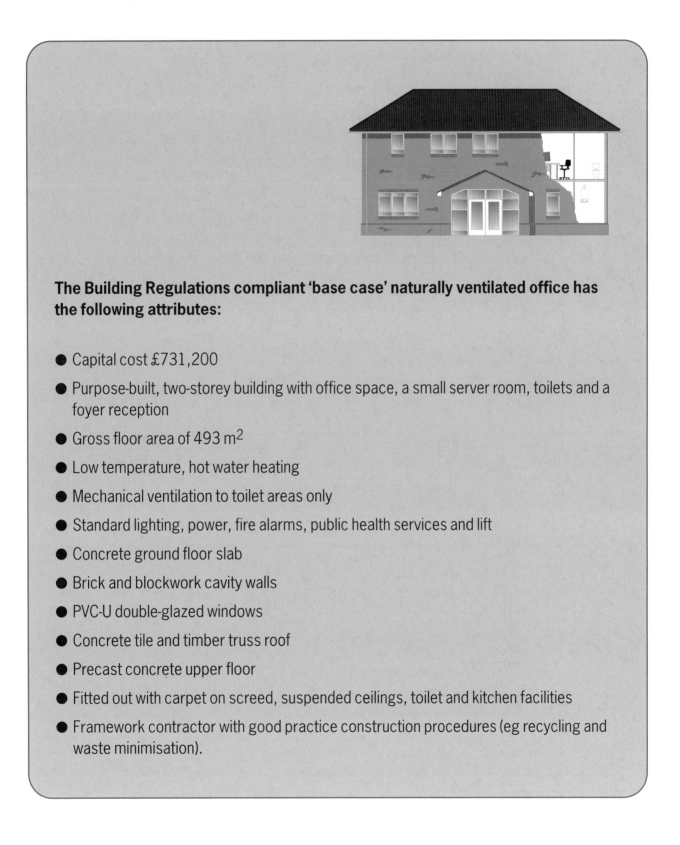

The Building Regulations compliant 'base case' naturally ventilated office has the following attributes:

- Capital cost £731,200
- Purpose-built, two-storey building with office space, a small server room, toilets and a foyer reception
- Gross floor area of 493 m^2
- Low temperature, hot water heating
- Mechanical ventilation to toilet areas only
- Standard lighting, power, fire alarms, public health services and lift
- Concrete ground floor slab
- Brick and blockwork cavity walls
- PVC-U double-glazed windows
- Concrete tile and timber truss roof
- Precast concrete upper floor
- Fitted out with carpet on screed, suspended ceilings, toilet and kitchen facilities
- Framework contractor with good practice construction procedures (eg recycling and waste minimisation).

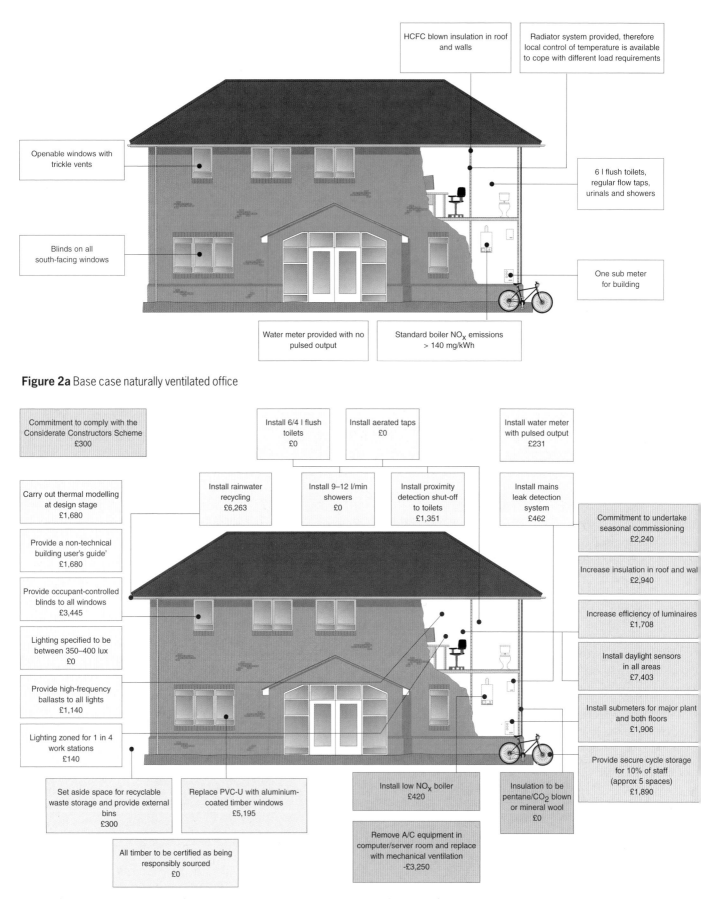

Figure 2a Base case naturally ventilated office

Figure 2b Additional capital costs for implementing more sustainable solutions (costs are for items implemented at the design stage, not retrofitted)

The environmental performance of this building was assessed with BREEAM Offices 2004. Figure 2b illustrates the environmental improvement measures assessed for BREEAM compliance and the additional costs associated with them compared with the base case naturally ventilated office. Table 2a lists the environmental improvement measures in order of best value.

Table 2a Additional capital costs for sustainable solutions along with the BREEAM points achievable. The measures are presented in order of best value

Description	Cost (£)	BREEAM points achieved[3]
Reducing CO$_2$ emissions from transport and operational energy		
Commissioning (all plant and equipment costs include testing and commissioning in line with best practice)	0	3.00
Install submeters for major plant and both floors	1,906	1.67
Commitment to undertake seasonal commissioning	2,240	1.50
Provide secure cycle facilities for 10% of staff (approximately 5 cycle spaces)	1,890	0.83
Increase insulation in roof and walls	2,940	0.83
Install daylight sensors in all areas and increase efficiency of luminaires	9,111	0.83
Reducing mains water consumption		
Install 6/4 litre flush toilets, aerated taps and 9–12 l/min showers	0	0.83
Install water meter with pulsed output (allowing future connection to a BMS)	231	0.83
Install mains leak detection system	462	0.83
Install proximity detection shut-off to toilet area	1,351	0.83
Install rainwater recycling	6,263	2.08
Minimising the impact on the local area		
Temporary site timber to be certified as being responsibly sourced	0	1.50
Commitment to improve on the standards laid down by the Considerate Constructors Scheme	300	3.00

Description	Cost (£)	BREEAM points achieved[3]
Reducing the impact of materials used		
Timber to be certified as being responsibly sourced	0	1.67
Set aside space for recyclable waste storage and provide external bins	300	0.83
Replace PVC-U windows with aluminium-coated timber windows	5,195	0.83
Reducing pollutants harmful to the atmosphere		
Remove A/C equipment in computer/server room and replace with mechanical ventilation	-3,250	1.25
Insulation to be pentane/CO$_2$ blown or mineral wool	0	1.25
Install low NO$_X$ boiler	420	3.75
Improving the indoor environment		
Lighting specified to be between 350–400 lux	0	1.00
Lighting zoned for 1 in 4 work stations to allow for local occupant control	140	1.00
Provide a non-technical building user's guide for occupants and facilities managers	1,680	1.50
Provide high-frequency ballasts to all lights	1,140	1.00
Carry out thermal modelling at design stage	1,680	1.00
Provide occupant-controlled blinds to all windows	3,445	1.00

[3] For a description of BREEAM points, credits and scoring see the Appendix on page 21.

Table 2b   Increases in capital costs to achieve Good, Very Good and Excellent BREEAM ratings in three locations

Location[2]	BREEAM score and rating for the base case naturally ventilated office	% increase in capital cost to achieve a Pass/Good/Very Good/Excellent			
		Pass	**Good**	**Very Good**	**Excellent**
Poor	25.4 (Pass)	−0.4	−0.3	2.0	—
Typical	39.7 (Pass)	—	−0.4	−0.3	3.4
Good	42.2 (Good)	—	—	−0.4	2.5

Thermal modelling demonstrated that it was possible to remove a small air-conditioning unit from the office server room without having an impact on the functionality of the room. This resulted in a substantial cost saving while securing several BREEAM credits through avoiding the use of refrigerants. Even in a poor location a BREEAM Good rating can be achieved while saving money, and in a typical or good location, a BREEAM Very Good can be achieved at no additional capital cost. It was not possible to gain an Excellent with the measures listed in Table 2a for a poor location but this could be achieved through considering the selection of site-related credits excluded from this study (see 'Our approach' on page 2). Figure 2c shows the incremental environmental improvements against additional cost, clearly demonstrating the large increase in BREEAM score achievable at no additional cost.

As with all the graphs in this study, sometimes the lines cross the rating boundaries at a lower % than is given in the tables. For an explanation of this, see page 6.

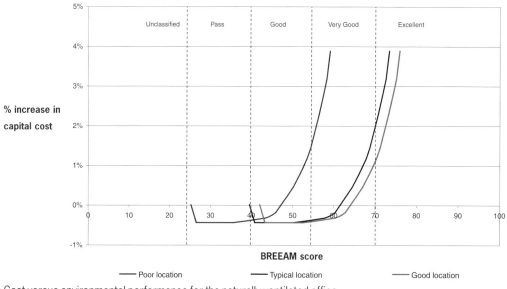

Figure 2c Cost versus environmental performance for the naturally ventilated office

Whole life implications

The energy- and water-saving measures identified for the naturally ventilated office provide predicted in-use cost savings of 17% and 71% respectively, throughout the life of the building.

---

2 For description of the three locations see 'Our appoach' on page 2.

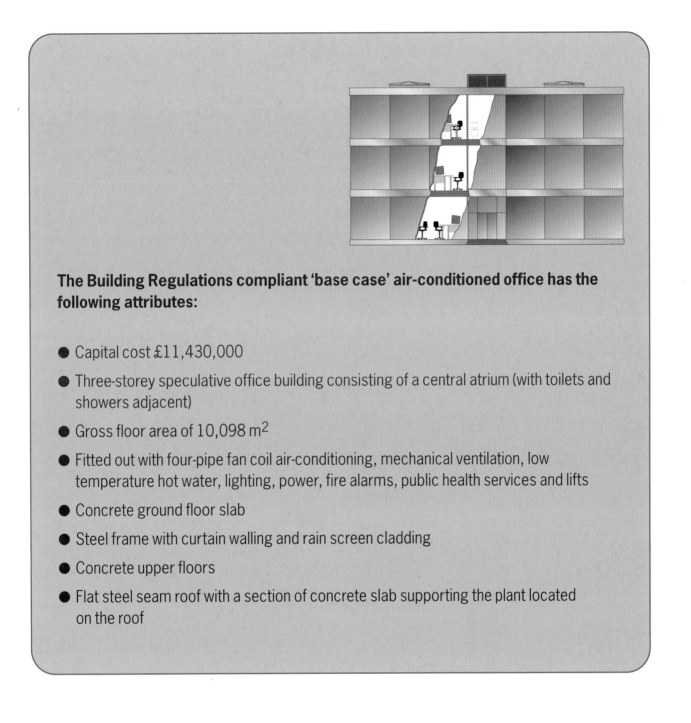

The Building Regulations compliant 'base case' air-conditioned office has the following attributes:

- Capital cost £11,430,000
- Three-storey speculative office building consisting of a central atrium (with toilets and showers adjacent)
- Gross floor area of 10,098 m^2
- Fitted out with four-pipe fan coil air-conditioning, mechanical ventilation, low temperature hot water, lighting, power, fire alarms, public health services and lifts
- Concrete ground floor slab
- Steel frame with curtain walling and rain screen cladding
- Concrete upper floors
- Flat steel seam roof with a section of concrete slab supporting the plant located on the roof

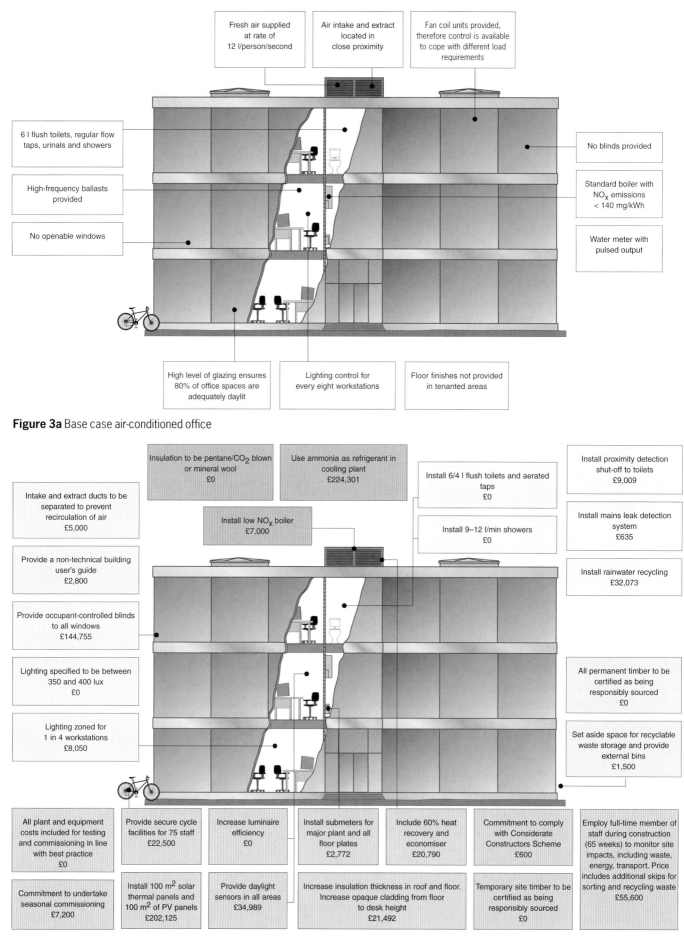

Figure 3a Base case air-conditioned office

The following labels appear in Figure 3a:

- Fresh air supplied at rate of 12 l/person/second
- Air intake and extract located in close proximity
- Fan coil units provided, therefore control is available to cope with different load requirements
- 6 l flush toilets, regular flow taps, urinals and showers
- No blinds provided
- High-frequency ballasts provided
- Standard boiler with NO$_x$ emissions < 140 mg/kWh
- No openable windows
- Water meter with pulsed output
- High level of glazing ensures 80% of office spaces are adequately daylit
- Lighting control for every eight workstations
- Floor finishes not provided in tenanted areas

The following labels appear in Figure 3b:

- Insulation to be pentane/CO_2 blown or mineral wool £0
- Use ammonia as refrigerant in cooling plant £224,301
- Install 6/4 l flush toilets and aerated taps £0
- Install proximity detection shut-off to toilets £9,009
- Intake and extract ducts to be separated to prevent recirculation of air £5,000
- Install low NO$_x$ boiler £7,000
- Install 9–12 l/min showers £0
- Install mains leak detection system £635
- Provide a non-technical building user's guide £2,800
- Install rainwater recycling £32,073
- Provide occupant-controlled blinds to all windows £144,755
- Lighting specified to be between 350 and 400 lux £0
- All permanent timber to be certified as being responsibly sourced £0
- Lighting zoned for 1 in 4 workstations £8,050
- Set aside space for recyclable waste storage and provide external bins £1,500
- All plant and equipment costs included for testing and commissioning in line with best practice £0
- Provide secure cycle facilities for 75 staff £22,500
- Increase luminaire efficiency £0
- Install submeters for major plant and all floor plates £2,772
- Include 60% heat recovery and economiser £20,790
- Commitment to comply with Considerate Constructors Scheme £600
- Employ full-time member of staff during construction (65 weeks) to monitor site impacts, including waste, energy, transport. Price includes additional skips for sorting and recycling waste £55,600
- Commitment to undertake seasonal commissioning £7,200
- Install 100 m² solar thermal panels and 100 m² of PV panels £202,125
- Provide daylight sensors in all areas £34,989
- Increase insulation thickness in roof and floor. Increase opaque cladding from floor to desk height £21,492
- Temporary site timber to be certified as being responsibly sourced £0

Figure 3b Additional capital costs for implementing more sustainable solutions (costs are for items implemented at the design stage, not retrofitted)

The environmental performance of this building was assessed with BREEAM Offices 2004. Figure 3b illustrates the environmental improvement measures assessed for BREEAM compliance and the additional costs associated with them compared with the base case air-conditoned office. Table 3a lists the environmental improvement measures in order of best value.

Table 3a Additional capital costs for sustainable solutions along with the BREEAM points achievable. The measures are presented in order of best value

Description	Cost (£)	BREEAM points achieved[3]
Reducing CO$_2$ emissions from transport and operational energy		
Commissioning (all plant and equipment costs include testing and commissioning in line with best practice)	0	3.00
Install submeters for major plant and all floor plates	2,772	1.67
Commitment to undertake seasonal commissioning	7,200	1.50
Include 60% heat recovery and economiser	20,790	0.83
Increase insulation in roof and floor. Increase opaque cladding from floor to desk height	21,492	0.83
Provide secure cycle facilities for 75 staff	22,500	0.83
Install daylight sensors in all areas and increase efficiency of luminaires	34,989	0.83
Install 100 m^2 solar thermal panels and 100 m^2 of PV panels	202,125	1.25

Description	Cost (£)	BREEAM points achieved[3]
Improving the indoor environment		
Lighting specified to be between 350–400 lux	0	1.00
Provide a non-technical building user's guide for occupants and facilities managers	2,800	1.50
Intake and extract ducts to be separated to prevent recirculation of air	5,000	1.00
Lighting zoned for 1 in 4 work stations to allow for local occupant control	8,050	1.00
Provide occupant-controlled blinds to all windows	144,755	1.00

Description	Cost (£)	BREEAM points achieved[3]
Reducing the impact of materials used		
All permanent timber to be certified as being responsibly sourced	0	1.67
Set aside space for recyclable waste and provide external bins	1,500	0.83

Description	Cost (£)	BREEAM points achieved[3]
Reducing mains water consumption		
Install 6/4 litre flush toilets, aerated taps and 9–12 l/min showers	0	0.83
Install mains leak detection system	635	0.83
Install proximity detection shut-off to toilets	9,009	0.83
Install rainwater recycling	32,073	2.08

Description	Cost (£)	BREEAM points achieved[3]
Reducing pollutants harmful to the atmosphere		
Insulation to be pentane/CO$_2$ blown or mineral wool	0	1.25
Install low NO$_x$ boiler	7,000	2.50
Use ammonia as a refrigerant in cooling plant	224,301	1.25

Description	Cost (£)	BREEAM points achieved[3]
Minimising the impact on the local area		
Temporary site timber to be certified as being responsibly sourced	0	1.50
Commitment to comply with the Considerate Constructors Scheme	600	1.50
Employ full-time member of staff during construction to monitor site impacts including waste, energy, transport	55,600	4.50

[3] For a description of BREEAM points, credits and scoring see the Appendix on page 21.

Table 3b   Increases in capital costs to achieve Good, Very Good and Excellent BREEAM ratings in three locations

Location[2]	BREEAM score and rating for the base case air-conditioned office	% increase in capital cost to achieve a Pass/Good/Very Good/Excellent			
		Pass	**Good**	**Very Good**	**Excellent**
Poor	20.3 (Unclassified)	0	0.2	5.7	—
Typical	34.6 (Pass)	—	0	0.2	7.0
Good	37.1 (Pass)	—	0	0.1	3.3

The use of refrigerants and the added use of energy in this air-conditioned office causes the base case scores to be lower than those of the naturally ventilated office. There are many other differences between the two buildings but they are not as significant in terms of environmental impact. Table 3b shows that in a typical and good location, a BREEAM Good rating can be achieved for no additional capital cost. Even for a location where a limited number of location credits are achievable, the Excellent rating can be achieved for just 7% additional cost. As for the previous case studies, in a poor location, an Excellent rating could not be achieved solely on the measures listed and more site-specific credits would need to be addressed. Figure 3c illustrates these results graphically, showing the large increase in BREEAM score achievable for little additional cost.

As with all the graphs in this study, sometimes the lines cross the rating boundaries at a lower % than is given in the tables. For an explanation of this, see page 6.

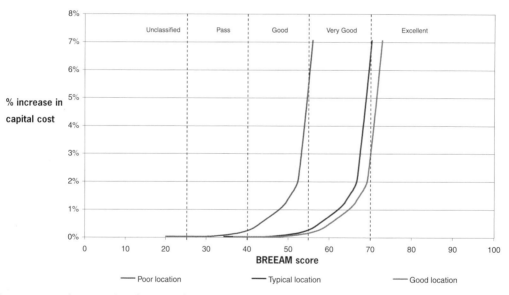

Figure 3c Cost versus environmental performance for the air-conditioned office

Whole life implications

The energy- and water-saving measures identified for the air-conditioned office provide predicted in-use cost savings of 26% and 55% respectively, throughout the life of the building.

[2] For description of the three locations see 'Our appoach' on page 2.

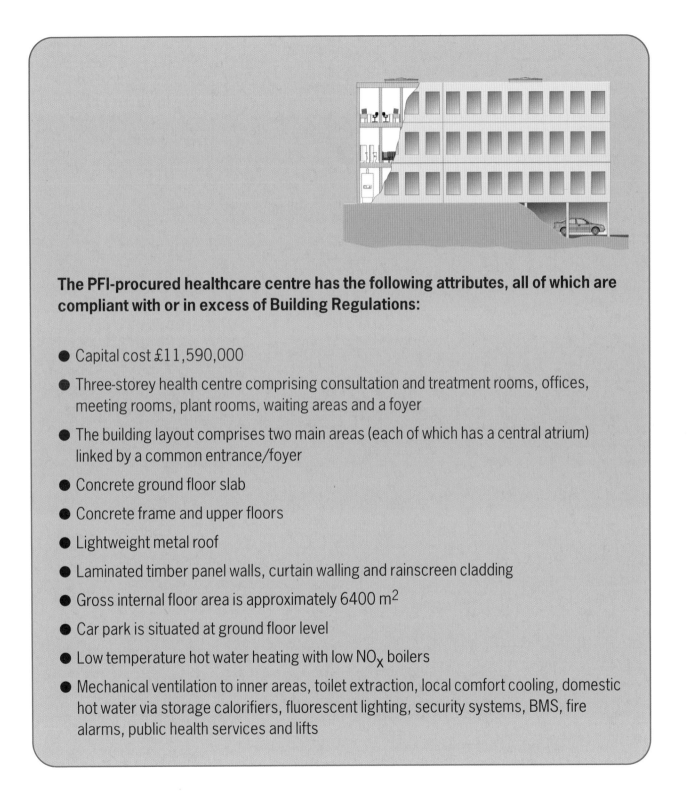

The PFI-procured healthcare centre has the following attributes, all of which are compliant with or in excess of Building Regulations:

- Capital cost £11,590,000

- Three-storey health centre comprising consultation and treatment rooms, offices, meeting rooms, plant rooms, waiting areas and a foyer

- The building layout comprises two main areas (each of which has a central atrium) linked by a common entrance/foyer

- Concrete ground floor slab

- Concrete frame and upper floors

- Lightweight metal roof

- Laminated timber panel walls, curtain walling and rainscreen cladding

- Gross internal floor area is approximately 6400 m^2

- Car park is situated at ground floor level

- Low temperature hot water heating with low NO_x boilers

- Mechanical ventilation to inner areas, toilet extraction, local comfort cooling, domestic hot water via storage calorifiers, fluorescent lighting, security systems, BMS, fire alarms, public health services and lifts

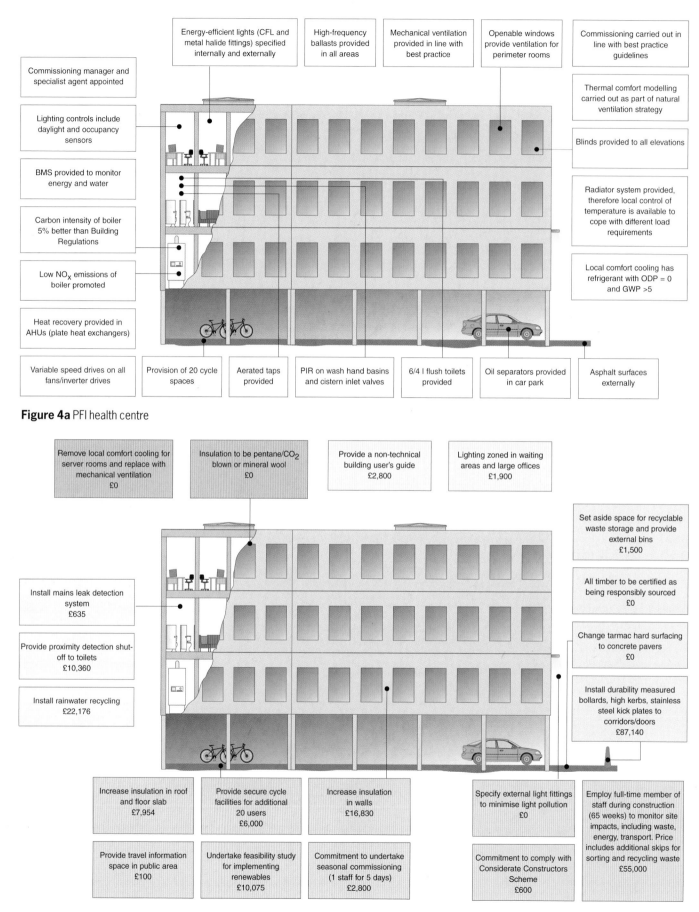

Figure 4a PFI health centre

Figure 4b Additional capital costs for implementing more sustainable solutions (costs are for items implemented at the design stage, not retrofitted)

The environmental performance of this building was assessed using Bespoke BREEAM (see Appendix A). Figure 4b illustrates the environmental improvement measures assessed for the Bespoke BREEAM compliance and the additional costs associated with them compared with the base case healthcare building. Table 4a lists the environmental improvement measures in order of best value for each of the environmental categories.

Table 4a Additional capital costs for sustainable solutions along with the BREEAM points achievable. The measures are presented in order of best value

Description	Cost (£)	BREEAM points achieved[3]
Reducing CO_2 emissions from transport and operational energy		
Provide travel information space in public area	100	0.81
Commitment to undertake seasonal commissioning (1 staff for 5 days)	2,800	1.25
Increase insulation in roof and floor slab	7,954	1.21
Provide secure cycle facilities for additional 20 users	6,000	0.81
Undertake feasibility study for implementing renewables	10,075	0.81
Increase insulation in walls	16,830	1.21

Description	Cost (£)	BREEAM points achieved[3]
Reducing pollutants harmful to the atmosphere		
Remove local comfort cooling for server rooms and replace with mechanical ventilation	0	4.02
Insulation to be pentane/CO_2 blown or mineral wool	0	1.07

Description	Cost (£)	BREEAM points achieved[3]
Improving the indoor environment		
Provide a non-technical building user's guide for occupants and facilities managers	2,800	1.25
Lighting zoned in waiting areas and large offices	1,900	0.63

Description	Cost (£)	BREEAM points achieved[3]
Reducing the impact of materials used		
All timber to be certified as being responsibly sourced	0	1.25
Change tarmac hard surfacing to concrete pavers	0	0.63
Set aside space for recyclable waste and provide external bins	1,500	0.63
Install durability measures, ie bollards, high kerbs, stainless steel kick plates to corridors/doors	87,140	0.63

Description	Cost (£)	BREEAM points achieved[3]
Minimising the impact on the local area		
Specify external light fittings to minimise light pollution	0	1.07
Commitment to comply with the Considerate Constructors Scheme	600	2.50
Employ full-time member of staff during construction to monitor site impacts including waste, energy, transport	55,000	2.50

Description	Cost (£)	BREEAM points achieved[3]
Reducing mains water consumption		
Install mains leak detection system	635	0.83
Install rainwater recycling	22,176	1.90
Install proximity detection shut-off to toilets	10,360	0.83

---

3 For a description of BREEAM points, credits and scoring see the Appendix on page 21.

Table 4b   Increases in capital costs to achieve Good, Very Good and Excellent BREEAM ratings in two locations

Location[2]	BREEAM score and rating for the base case healthcare facility	% increase in capital cost to achieve a Pass/Good/Very Good/Excellent			
		Pass	Good	Very Good	Excellent
Typical	44.3 (Good)	—	—	0	1.9
Good	48.4 (Good)	—	—	0	0.6

The final case study assesses a building procured by the Private Finance Initiative (PFI). It appears that this method of procurement has a positive influence on the building's environmental performance and this is probably due to the developer/contractor's long-term interest in building operations. For example, the base case contains water- and energy-efficient appliances/ services and a building management/monitoring system. A healthcare facility, by its nature, is likely to be located in an area close to the community with at least some access to public transport. Therefore, an analysis of this building in a poor location has not been included as it would be unrepresentative.

As the PFI health centre automatically achieves a Good BREEAM rating in both the typical and good locations, the 1.9% increase on the base case cost to achieve an Excellent in a typical location is lower than for the other case studies due to a higher starting point. This is displayed graphically in Figure 4c.

> As with all the graphs in this study, sometimes the lines cross the rating boundaries at a lower % than is given in the tables. For an explanation of this, see page 6.

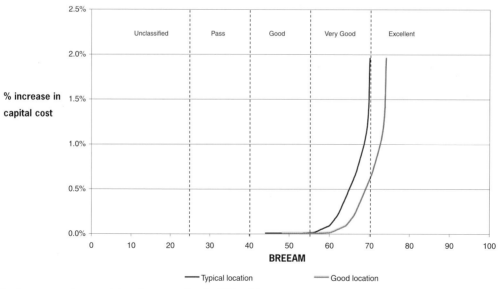

Figure 3c Cost versus environmental performance for the PFI health centre

> **Whole life implications**
> The energy- and water-saving measures identified for the health centre provide predicted in-use cost savings of 3% and 10% respectively, throughout the life of the building.

2 For description of the three locations see 'Our appoach' on page 2.

discussion

This study illustrates that many sustainability measures can be implemented at little cost and a limited number of items are available at no additional cost and may even offer a cost saving.

Other broadly applicable findings are as follows.

- Development location and site conditions have a major impact on the costs associated with achieving higher (ie Very Good or Excellent) BREEAM/EcoHomes ratings. Any estimates of the cost of achieving a BREEAM rating should include a thorough review of site conditions.
- Effective management of the development process is critical to ensuring that all low cost options are identified and achieved. Costs can rapidly increase once all the low cost options have been implemented/exhausted. To enable environmental performance to be maximised at lowest cost, sustainability must be considered at the earliest possible stage of the development process and every effort made to incorporate all the possible low cost items.
- The case studies in this report assume that solutions are implemented at the design stage, rather than retrofitted. Identified low or no-cost options include:
  - ❏ specifying water-efficient appliances,
  - ❏ ensuring all timber is procured from appropriate certified sources,
  - ❏ committing to good construction practice (such as through the Considerate Constructors Scheme),
  - ❏ providing low energy lighting,
  - ❏ enhancing thermal performance through increased insulation levels.

- Air-conditioning should be avoided where possible. Not only does this help to reduce energy use and associated CO_2 emissions, but the use of a passive or mechanical system also avoids the need for expensive refrigerants (particularly those with low global warming potential) and refrigerant-monitoring systems.
- Providing technologies such as photovoltaic panels or rainwater harvesting is still relatively expensive compared with most other sustainability options.
- While outside the scope of this study, good controls and commissioning can lead to significantly lower energy consumption making it more likely that design aspirations will be achieved.

Purchasing or procuring an energy-efficient building can result in significant cost savings from energy bills and if the mains water consumption is metered, implementing low water use facilities will also have a positive effect on reducing operational expenditure. This is a benefit to owner occupiers, but can also be positive when selling or letting an office building.

> Long-term energy-saving measures are also an important consideration, given the imminent implementation of the new Energy Performance of Buildings Directive. This Directive will require all new buildings, public buildings and buildings being let or sold to have an energy rating prominently displayed. This will increase the value of including energy-saving items in any new building. While initially being applied to non-domestic buildings, the Directive will require energy certification of residential buildings at the point of sale in the near future.

The case studies in this report indicate the nature of costs involved in procuring a building with Good, Very Good and Excellent EcoHomes/BREEAM ratings. It is important to remember, however, that specific costs cannot be transferred directly to any site or situation. Each project must be considered in terms of its own opportunities. Items to consider include:
● construction programme,
● stage of development,
● site location, and
● the project team's standard procedures and specifications.

Ensuring that sustainability is considered at the appropriate stage will minimise costs and maximise environmental performance. The case studies in this report illustrate a number of items as being cost neutral, though this may not be the case if the sustainability advice is received too late. Figure 5 has been produced to assist with optimising project sustainability performance. This schematic diagram outlines a range of sustainability issues and ideally when these should be considered by the design team, planner, contractors, owners/occupiers and other members of the project team.

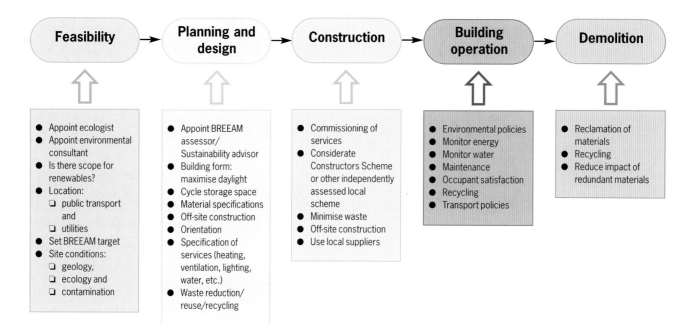

Figure 5 Sustainability issues to consider during the lifetime of a building

BRE's Environmental Assessment Method (BREEAM) is used to measure a building's environmental performance. The main objectives of the scheme are:
● To differentiate developments that meet higher levels of environmental performance from the rest of the market.
● To encourage, reward and promote best environmental practice.
● To raise awareness of the benefits of building to best environmental practice standards.

BREEAM and EcoHomes (BREEAM for houses) consider broad environmental concerns such as climate change, resource use and impacts on wildlife, and balances these against the needs for a high quality, safe and healthy internal environment. They identify those projects that improve environmental performance through good management, design and specification, and credits are awarded where specific performance levels are achieved in each category. The credits are then weighted according to the relative importance of the environmental issue that they address to give a point score. The weightings were established through a large stakeholder consultation and are revised periodically as priorities change. The following ratings are awarded on a scale of achievement: Pass, Good, Very Good and Excellent, based on the scores given in Table 5.

Where the standard assessment methods are not applicable, the Bespoke BREEAM scheme can be applied. In these cases, the BREEAM criteria are specifically adapted to account for the functions within the building. All the issues within the Bespoke BREEAM assessment are based on the existing BREEAM methods, but benchmarks may be altered to suit the building's use.

In this study, the most recent versions of BREEAM that were available at the start of this study were used:
● BREEAM Offices 2004,
● EcoHomes 2003, and
● Bespoke BREEAM 2004.

For more information on BREEAM and EcoHomes visit: www.breeam.org.uk

Table 5   BREEAM and EcoHomes scoring

	BREEAM score	**EcoHomes score**
Unclassified	< 25	< 36
Pass	≥ 25 – < 40	≥ 36 – < 48
Good	≥ 40 – < 55	≥ 48 – < 60
Very good	≥ 55 – < 70	≥ 60 – < 70
Excellent	≥ 70	≥ 70